The True Wireless
By Nikola Tesla
(1919)

Contents

Introduction
by George Trinkaus....... i – iv

The True Wireless
by Nikola Tesla............. 1 – 13

Explanatory Notes...... 14

For more information... 15

Introduction

by George Trinkaus

Tesla wrote *The True Wireless* back in 1919 because he wanted to knock the prevailing orthodoxy of radio theory into a cocked hat. I am republishing the piece in 1998 because that orthodoxy still prevails and because I agree with Tesla that into a cocked hat is where this persistent and annoying body of dogma should be knocked.

How does radio really work? What responsible scientific explanations can we propose for the mysteries of wireless? What are the true dynamics of propagation in as much as we can ever know them?

Unfortunately, early on in the development of radio, explanations were proposed, and, as is the sad habit of academic science, some of these conjectures readily solidified into the official theory.

Does modern science have no patience with mystery? In 1901 long-distance wireless was achieved. In the very next year, a respected scientist, Oliver Heavyside, suggested this might be done by signals bouncing off of ionized layers presumed to exist a hundred or so miles up. Perhaps because it was consistent with the already accepted notion that radio behaved like light rays that can be reflected, the Heavyside layer (later called the ionosphere) had instant academic appeal, and, despite the protest of thinkers with other ideas about how radio worked, the theory of ionospheric propagation became a fixture in the official theory, enabling radiomen through the century to speak glibly of "skip," whether such a phenomenon demonstrably exists or not. Tesla said, "There is no Heavyside layer, or, if it exists, it is of no effect."

Here in a nutshell is the official theory of wireless propagation in the ham bands as abstracted from the 1996 edition of the American Radio Relay League's *Radio Amateur's Handbook:*

Radio is radiation. Radio is waves. Radio waves behave like light rays. Unobstructed, as in the "vacuum" of space, radio moves in straight lines. Like light it can be refracted, reflected, scattered, or absorbed and travels at 186,000 miles per second. The whole model of orthodox radio is demystified through this analogy to the observable behavior of light.

The layers of the so-called ionosphere vary in density, and, hence refractivity. Under the influence of sunlight, there are day-night fluctuations and 11-year sun-spot cycles which enhance or diminish ionospheric reflection and hence the ability of short-wave radio to propagate long-distance in this fashion.

Radio is assumed to be an aerial phenomenon. Although sometimes a surface-travelling "ground wave" is acknowledged to account for propagation at lower frequencies, conduction through the earth, which Tesla proposes, is implied to be an impossibility in the orthodoxy. To Tesla, radio propagation *is* conduction.

Tesla says here that orthodox Hertz radio theory is "one of the most remarkable and inexplicable aberrations of the scientific mind which has ever been recorded in history" and that it "has stifled creative effort in the wireless art and retarded it for twenty-five years."

Orthodoxy drives out speculation. We can be grateful to Tesla for his attempt to smash this orthodoxy and for his giving us, if we go with him, a new beginning in this neglected science. Liberated, we can proceed to wonder again at the mysteries, to speculate, experiment, even have some fun.

Mostly for the last reason, I built a sensitive field-strength meter, connected it between an antenna and ground, tuned it to a strong local AM broadcast signal, and sat back and watched. I used a large-face microammeter that could be read from fifteen feet away. Set on a worktable near the foot of my bed, it became a source of entertainment, my etheric TV.

It was essentially the simplest crystal receiver, with the 0-to-200 microammeter in the circuit where the earphones normally go. For amplification, Tesla-style resonant amplification, I connected a big coil with a parallel variable capacitor as a tank tuner

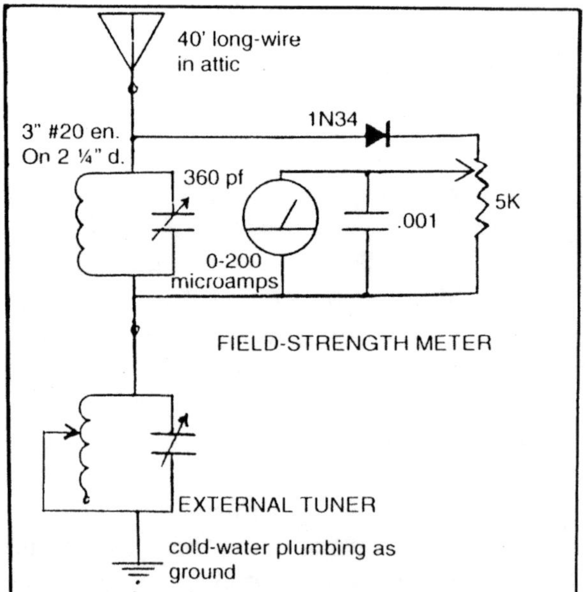

signal-strength experiment

between field-strength meter and ground. I had also tried it between meter and antenna but achieved greater effect in the ground circuit, which was suggestive. (If you connect your AM or shortwave receiver's antenna terminal to ground instead of to an aerial, you commit a Hertzian heresy, but how do you explain the improved reception?) As measured on my vacuum-tube voltmeter, 1.5 volts or more can be coaxed out of ground by tuning in the above fashion. Typical of such an elementary crystal set in the city, reception was dominated by one or another of two powerful local stations. I tuned the system to the strongest meter reading, which gave me 50-kilowatt KEX (Portland, Oregon) at 1190 kilocycles, whose transmitter was less than five miles away. The meter's needle swung to mid-range and held, twitching to the signal's modulation (which, surprisingly, was less than ten percent of its deflection).

Now one would expect that a 50 kw signal from only five miles away would be of steady intensity, but variations over a period of days could be as dramatic as 50 microamps. It was interesting to try to correlate variations with weather conditions, day and night, lunar cycles, etc., but none of these speculations was conclusive. What was conclusive was that nothing in conventional propagation theory accounted for what I was seeing.

Hertzian theory does allow for propagation at this relatively low frequency by means of the so-called ground wave; thus any variations in the alleged ionosphere would not be a factor, nor would it be anyway at this short distance.

So here is an instance of how the orthodoxy is empty. And here is an invitation to speculate anew. What is the true nature of the radio-conductive medium? I don't know. You don't know. We can only guess. Evidently the medium is rich in influences. The atmosphere, the earth (let's call it all the electrosphere) may be coursed by currents, not all electrical necessarily but also of other "unusual energies," including perhaps those known to the dowser. These energies might exercise attenuating and even enhancing influences on the radio energies imposed by man. So let's hypothesize that the medium could be a veritable soup of churning energetic activity that can affect a signal. Let's appreciate our ignorance, be humble, and refrain from smug assertions, dear ARRL and its graduates.

By June 1919 when Hugo Gernsbach published Tesla's "The True Wireless" in his *Electrical Experimenter*, Tesla was well on his way to being rendered invisible by the media, and he must have been grateful for this opportunity to express himself publicly. In earlier years, the media had romanced Tesla to the public as the "electrical genius" who had invented the alternating-current system, and he had enjoyed great access to the press, but when Tesla reinvented his own system, after it had been established industrially as the official system, his voice had to be silenced.

By 1921 Tesla was unpublishable, officially at least. In the few years prior to that, the *Experimenter* published Tesla profusely. Tesla's publication of "The True Wireless" in the *Experimenter* would appear to have been granted grudgingly, though, considering the way in which the article was laid out by the magazine's graphic people. Clarity is sacrificed to "pack," for what you read here in 11-point type over thirteen pages, the *Experimenter* published in 8-point over four. Because of this tight economy, and because of the limitations of metal-type layout in the letterpress era, the illustrations in the original are often placed remotely from the relevant text. Moreover, many of the original illustrations are tiny (even by High Voltage Press standards). All this makes the reading of the original a drag. I trust that my drudgeries in resetting and redesigning this work make your experience a more pleasant one.

Tesla, in this writing, aware that in respect to the establishment he had nothing to lose, indulges a candor that he admits he could not indulge previously "on the ground that such idle and far-fetched speculations would injure me in the opinion of conservative businessmen."

Tesla's challenge to the radio theory he blames on Hertz comes from the same place as Tesla's challenge to the orthodoxy of electric power transmission. The same electrics are involved. Tesla's far-fetched, speculative wireless is based on the electrics which allow for the phenomenon of the single wire without return. Tesla is determined that this radical novelty be comprehended by his brain-washed readers and he resorts to illustrations using crude hydraulic analogs (Figures 4 - 6). If you have seen a hot Tesla coil in action, then you've been shown a more poetic example of Tesla's electrics in the corona, sparks, and streamers that leap from the terminal, eagerly seeking conduction in the ambient medium. The single-wire-without-return phenomenon is at the heart of Tesla's radio and wireless power.

Done Tesla's way, the power of radio is "unlimited." But, says Tesla, "the experts are blind to the possibilities." *The True Wireless* is Tesla's lament on this theme. The tone suggests the frustration of the inventor who has been standing by mutely over the years seeing his radio used but at the same time misused and the phenomenon systematically misexplained even to its practitioners.

No ionosphere. Not an aerial phenomenon. Antenna current irrelevant. Parallel antennas unnecessary. Mountains irrelevant. Short waves ineffective. What heresies! Where would the average ham be without these precepts? His license exam requires at least a rote grasp of the orthodoxy; his social milieu takes it for granted.

But the old dogma does not serve us, says Tesla. It is the theory of an impotent radio. "The transmitter generates several systems of waves all of which, except one, are useless," he has said elsewhere, suggesting that a conventional transmitter works only inadvertently. Evidently, the active element even in conventional "Hertzian" radio is Tesla's electric conduction, but this goes unacknowledged. The result is what he has called an "unfit apparatus" for the job. Tesla's radio is so potent that "transmission is absolutely unlimited as to terrestrial distance and the amount of energy," which means energy sufficient not only for communications but for the lighting of cities and the driving of the machinery of industry.

This, says Tesla, is the true wireless.

The True Wireless
By Nikola Tesla

Ever since the announcement of Maxwell's electromagnetic theory, scientific investigators all the world over had been bent on its experimental verification. They were convinced it could be done and lived in an atmosphere of eager expectancy, unusually favorable to the reception of any evidence to this end. No wonder then that the publication of Dr. Heinrich Hertz's results caused a thrill as had scarcely ever been experienced before.

At that time I was in the middle of pressing work in connection with the commercial introduction of my system of power transmission, but, nevertheless, caught the fire of the enthusiasm and fairly burned with desire to behold the miracle with my own eyes. Accordingly, as soon as I had freed myself of these imperative duties and resumed research work in my laboratory on Grand Street, New York, I began, parallel with high-frequency alternators, the construction of several forms of apparatus with the object of exploring the field opened up by Dr. Hertz.

Recognizing the limitations of the devices he had employed, I concentrated my attention on the production of a powerful induction coil but made no notable progress until a happy inspiration led to the invention of the oscillation transformer. In the latter part of 1891 I was already so far advanced in the development of this new principle that I had at my disposal means vastly superior to those of the German physicist. All my previous efforts with Rhumkorf coils had left me unconvinced, and in order to settle my doubts I went over the whole ground once more, very carefully, with these improved appliances. Similar phenomena were noted, greatly magnified in intensity, but they were susceptible of a different and more plausible explanation.

I considered this so important that in 1892 I went to Bonn, Germany, to confer with Dr. Hertz in regard to my observations. He seemed disappointed to such a degree that I regretted my trip and parted from him sorrowfully. During the succeeding years I made numerous experiments with the same object, but the results were uniformly negative. In 1900, however, after I had evolved a wireless transmitter which enabled me to obtain electromagnetic activities of many millions of horsepower, I made a last desperate attempt to prove that the disturbances emanating from the oscillator were ether vibrations akin to light, but met again with utter failure. For more than eighteen years I have been reading treatises, reports of scientific transactions, and articles on Hertz wave telegraphy, to keep myself informed, but they have always impressed me like works of fiction.

The history of science shows that theories are perishable. With every new truth that is revealed we get a better understanding of Nature, and our conceptions and views are modified. Dr. Hertz did not discover a new principle. He merely gave material support to a hypothesis which had been long ago formulated. It was a perfectly well established fact that a circuit, traversed by a periodic current, emitted some kind of space waves, but we were in ignorance as to their character. He apparently gave an experimental proof they were transversal vibrations in the ether. Most people look upon this as his great accomplishment. To my mind it seems that his immortal merit was not so much in this as in the focusing of the investigator's attention on the processes taking place in the ambient medium. The Hertz wave theory, by its fascinating hold on the imagination, has stifled creative effort in the wireless art and retarded it for twenty-five years. But, on the other hand, it is impossible to over-estimate the beneficial effects of the powerful stimulus it has given in many directions.

As regards signaling without wires, the applications of these radiations for the purpose was quite obvious. When Dr. Hertz was asked whether such a system would be of practical value, he did not think so, and he was correct in his forecast. The best that might have been expected was a method of communication similar to the heliographic and subject to the same or even greater limitations.

In the spring of 1891 I gave my demonstrations with a high-frequency machine before the American Institute of Electrical Engineers at Columbia College, which laid the foundation to a new and far more promising departure. Although the laws of electrical resonance were well known at that time and my lamented friend, Dr. John Hopkinson, had even indicated their specific application to an alternator in the Proceedings of the Institute of Electrical Engineers, London, Nov. 13, 1889, nothing had been done towards the practical use of this knowledge and it is probable that those experiments of mine were the first public exhibition with resonant circuits, more particularly of high frequency. While the spontaneous success of my lecture was due to spectacular features, its chief import was in showing that all kinds of devices could be operated through a single wire without return. This was the initial step in the evolution of my wireless system. The idea presented itself to me that it might be possible, under observance of proper conditions of resonance, to transmit electric energy through the earth, thus dispensing with all artificial conductors. Anyone who might wish to examine impartially the merit of that early suggestion must not view it in the light of present day science. I only need to say that as late as 1893, when I had prepared an elaborate chapter on my wireless system, dwelling on its various instrumentalities and future prospects, Mr. Joseph Wetzler and other friends of mine emphatically protested against its publication on the ground that such idle and far-fetched speculations would injure me in the opinion of conservative businessmen. So it came that only a small part of what I had intended to say was embodied in my address of that year before the Franklin Institute and the National Electric Light Association under the chapter "On Electrical Resonance." This little salvage from the wreak has earned me the title of "Father of the Wireless" from many well-disposed fellow workers, rather than the invention of scores of appliances which have brought wireless transmission within the reach of every young amateur and which, in a time not distant, will lead to undertakings overshadowing in magnitude and importance all past achievements of the engineer.

The popular impression is that my wireless work was begun in 1893, but as a matter of fact, I spent the two preceding years in investigations, employing forms of apparatus, some of which were almost like those of today. It was clear to me from the very start that the successful consummation could only be brought about by a number of radical improvements. Suitable high-frequency generators and electrical oscillators had to first to be produced. The energy of these had to be transformed in effective transmitters and collected at a distance in proper receivers. Such a system would be manifestly circumscribed in its usefulness if all extraneous interference were not prevented and exclusiveness secured. In time, however, I recognized that devices of this kind, to be most effective and efficient, should be designed with due regard to the physical properties of this planet and the electrical conditions obtaining in same. I will briefly touch upon the salient advances as they were made in the gradual development of the system. The high-frequency alternator employed in my first demonstrations is illustrated in Figure 1. It comprised a field ring, with 384 pole projections and a disk armature with coils wound in one single layer which were connected in various ways according to requirements. It was an excellent machine for experimental purposes, furnishing sinusoidal currents of from 10,000 to 20,000 cycles per second. The output was comparatively large, due to the fact that as much as 30 amperes per square millimeter could be passed through the coil without injury.

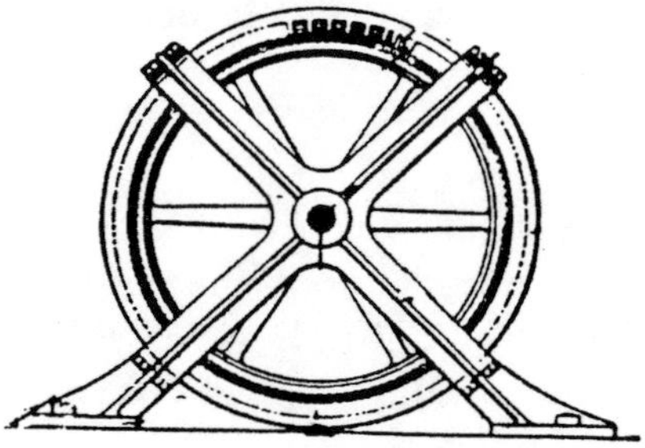

1. high-frequency alternator

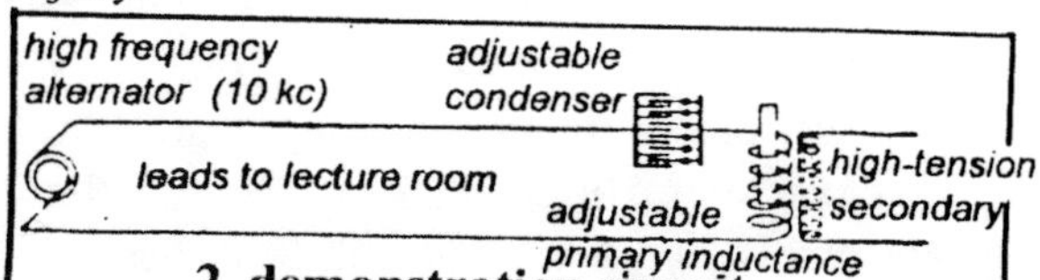

2. demonstration circuit

The diagram in Fig. 2 shows the circuit arrangements as used in my lecture. Resonant conditions were maintained by means of a condenser subdivided into small sections, the finer adjustments being effected by a movable iron core within an inductance coil. Loosely linked with the latter was a high tension secondary which was tuned to the primary.

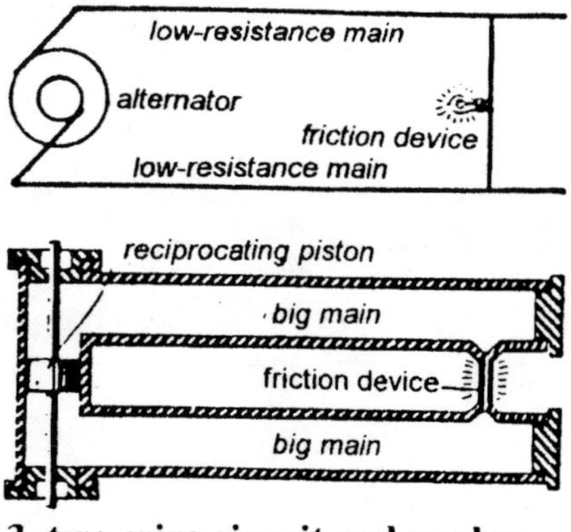

3. two-wire circuit and analog

The operation of devices through a single wire without return was puzzling at first because of its novelty, but can be readily explained by suitable analogs. For this purpose reference is made to Figs. 3 and 4.

In the former the low resistance electric conductors are represented by pipes of large section, the oscillator by an oscillating piston and the filament of an incandescent lamp by a minute channel connecting the pipes. It will be clear from a glance at the diagram that very slight excursions of the piston would cause the fluid to rush with high velocity through the small channel and that virtually all the energy of movement would be transformed into heat by friction, similarly to that of an electric current in the lamp filament.

The second diagram will now be self-explanatory. Corresponding to the terminal capacity of the electric system, an elastic reservoir is employed which dispenses with the necessity of a return pipe. As the piston oscillates, the bag

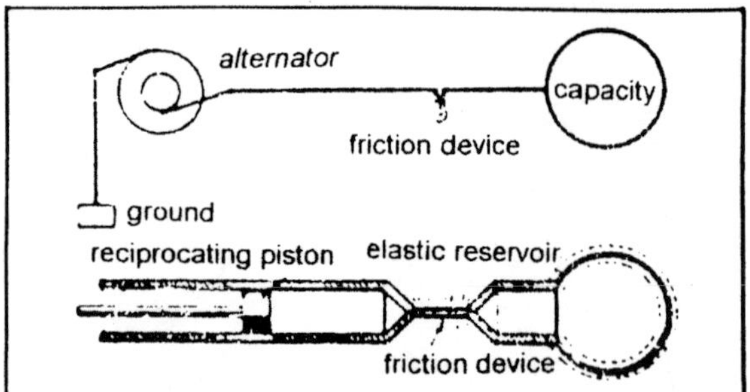

4. single wire without return and analog

expands and contracts, and the fluid is made to surge through the restricted passage with great speed, this resulting in the generation of heat in the incandescent lamp. Theoretically considered, the efficiency of conversion of energy should be the same in both cases.

Granted, then, that an economic system of power transmission through a single wire is practicable, the question arises how to collect the energy in the receivers. With this object attention is called to Fig. 5, in which a conductor is shown excited by an oscillator joined to it at one end. Evidently, as the periodic impulses pass through the wire, differences of potential will be created along the same as well as in right angles to it in the surrounding medium and either of these may be usefully applied. Thus at *a*, a circuit comprising an inductance and a capacity is resonantly excited in the transverse, and at *b*, in the longitudinal sense. At *c*, energy is collected in a circuit parallel to the conductor

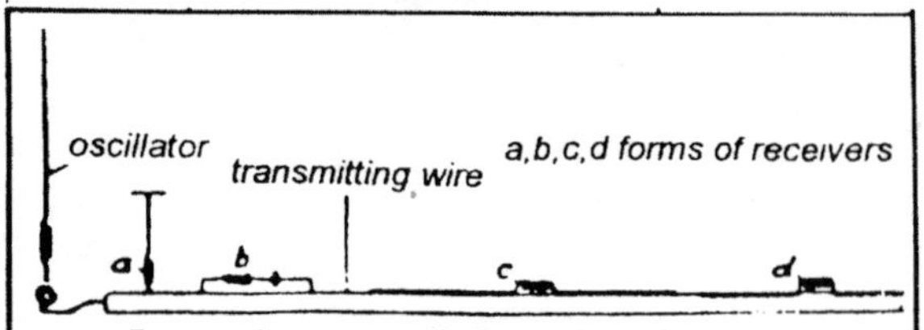

5. various receiving circuits

but not in contact with it, and again at *d*, in a circuit which is partly sunk into the conductor and may be, or not, electrically connected to same. It is important to keep these typical dispositions in mind, for, however the distant actions of the oscillator might be modified through the immense extent of the globe the principles involved are the same.

Consider now the effect of such a conductor of vast dimensions on a circuit exciting it. The upper diagram of Fig. 6 illustrates a familiar oscillating system comprising a straight rod of self-inductance $2L$ with small center capacities cc and a node in the center. In the lower dia-

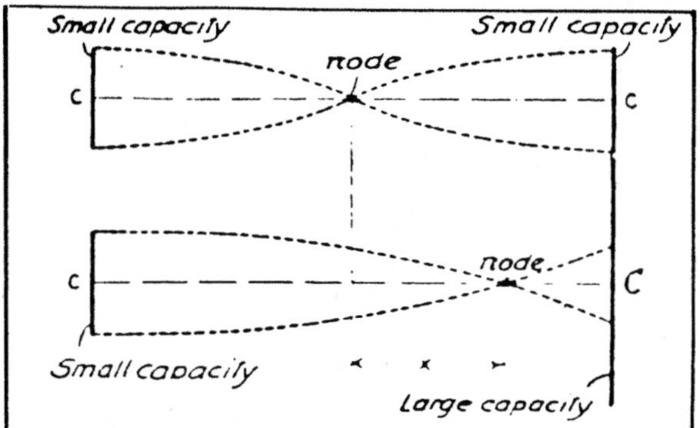

6. Effect of large capacity at one end

gram of the figure a large capacity C is attached to the rod at one end with the result of shifting the node to the right, through a distance corresponding to self- inductance X. As both parts of the system on either side vibrate at the same rate, we have evidently, (L + X), c = (L - X)C from which

$$X = L\,\frac{C - c}{C + c}$$

When the capacity C becomes commensurate to that of the earth, X approximates L, in other words, the node is close to the ground connection. *The exact determination of its position is very important in the calculation of certain terrestrial electrical and geodetic data* and I have devised special means with this purpose in view.

My original plan of transmitting energy without wires is shown in the upper diagram of Fig. 7, while the lower one illustrates its mechanical analog, first published in my article in the Century magazine of June 1900. An alternator, preferably of high tension, has one of its terminals connected to the ground and the other to an elevated capacity and impresses its oscillations upon the earth. At a distant point a receiving circuit, one of its terminals likewise connected to ground and the other to an elevated capacity, collects some of the energy and actuates a suitable device. I suggested a multiplication of such units in order to intensify the effects, an idea that may yet prove valuable. In the analog two tuning forks are provided, one at the sending and the other at the receiving station, each having attached to its lower prong a piston fitting in a cylinder. The two cylinders communi-

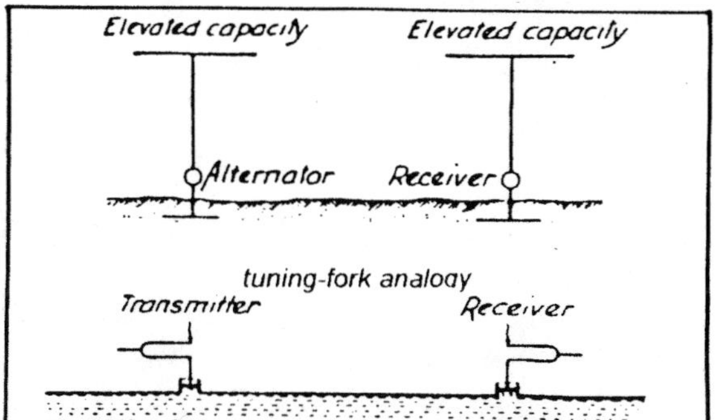

7. Wireless power through the earth

cate with a large elastic reservoir filled with an incompressible fluid. The vibrations transmitted to either of the tuning forks excite them by resonance and, through electrical contacts or otherwise, bring about the desired result. This, I may say, was not a mere mechanical illustration, but a simple representation of my apparatus for submarine signaling perfected by me in 1892, but not appreciated at that time, although more efficient than the instruments now in use.

The electric diagram in Fig. 7, which was reproduced from my lecture, was meant only for the exposition of the principle.The arrangement, as I described it in detail is shown in Fig. 8. In this case an alternator energizes the primary of a transformer, the high-tension secondary of which is connected to the ground and an elevated capacity and tuned to the impressed oscillations. The receiving circuit consists of an in-

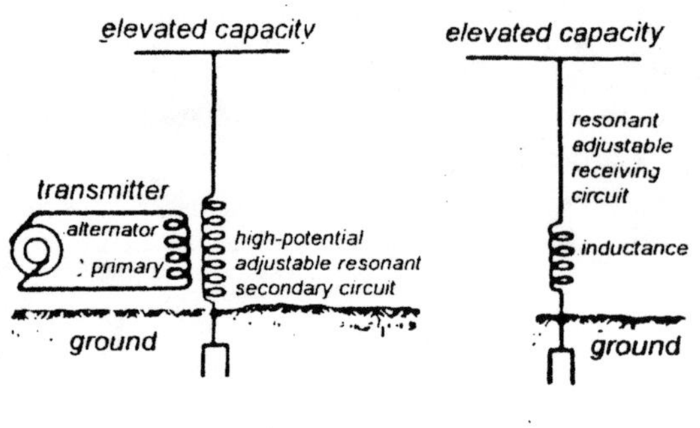

8. Wireless power schematic

ductance connected to the ground and to an elevated terminal without break and is resonantly responsive to the transmitted oscillations. A specific form of receiving device was not mentioned, but I had in mind to transform the received currents and thus make their volume and tension suitable for any purpose. This, in substance, is the system of today, and I am not aware of a single authenticated instance of successful transmission at considerable distance by different instrumentalities. It might perhaps be clear to those who have perused my first description of these improvements that, besides making known new and efficient types of apparatus, I gave the world a wireless system of potentialities far beyond anything before conceived.

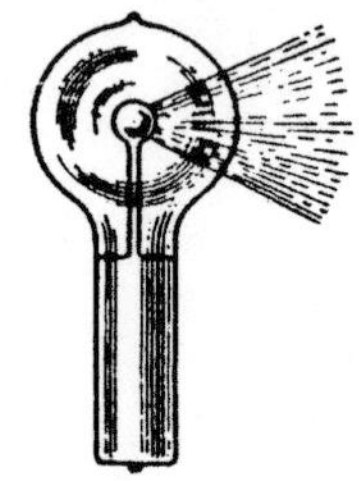

9. rotating brush

I made explicit and repeated statements that I contemplated transmission, absolutely unlimited as to terrestrial distance and amount of energy. But though I have overcome all obstacles which seemed in the beginning unsurmountable and found elegant solutions of all the problems which confronted me, experts are still blind to the possibilities which are within easy attainment.

My confidence that a signal could be easily flashed around the globe was strengthened through the discovery of the "rotating brush,"a wonderful phenomenon which I have fully described in my address before the Institution of Electrical Engineers, London, in 1892, and which is illustrated in Fig. 9. This is undoubtedly the most delicate wireless detector known, but for a long time it was hard to produce and to maintain such a sensitive state. These difficulties do not exist now, and I am looking to valuable applications of this device, particularly in connection with a high-speed photographic method, which I suggested, in wireless, as well as in wire, transmission.

Possibly the most important advances during the following three or four years were my system of concatenated tuned circuits and methods

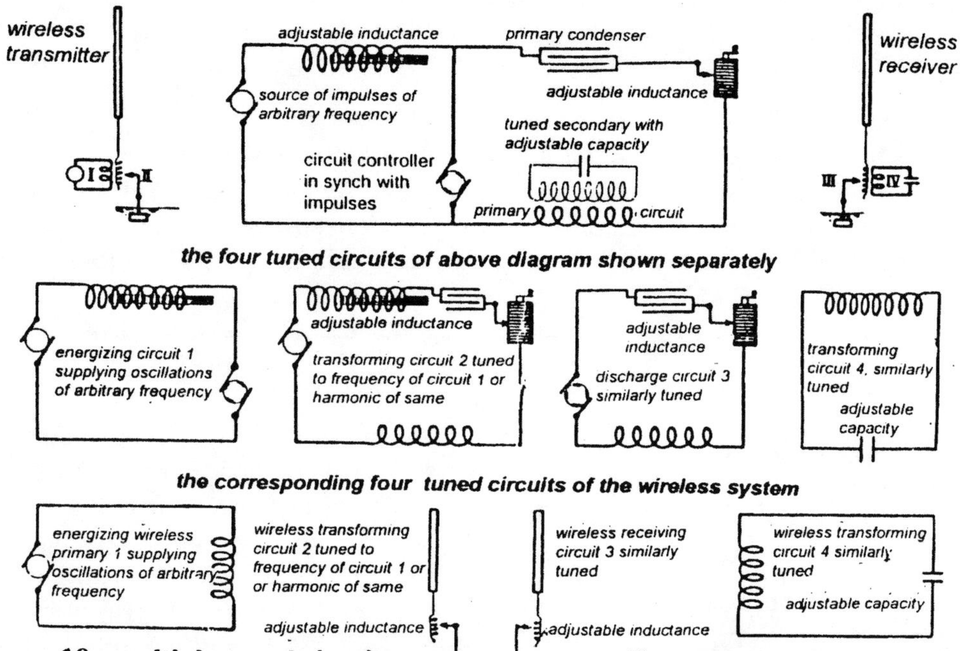

10. multiple tuned circuits ("concatenated")

of regulation, now universally adopted. The intimate bearing of these inventions on the development of the wireless art will appear from Fig. 10, which illustrates an arrangement described in my U.S. Patent No. 568178 of September 22, 1896, and corresponding depositions of wireless apparatus. The captions of individual diagrams are thought sufficiently explicit to dispense with further comment. I will merely remark that in this early record, in addition to indicating how any number of resonant circuits may be linked and regulated, I have shown the advantage of the proper timing of primary impulses and use of harmonics. In a farcical wireless suit in London, some engineers, reckless of their reputation, have claimed that my circuits were not at all atuned; they asserted that I had looked upon resonance as a wild and untamable beast.

It will be of interest to compare my system as first described in a Belgian patent of 1897 with the Hertz wave system of that period. The significant differences between them will be observed at a glance. The first enables us to transmit economically energy to any distance and is of inestimable value: the latter is capable of a radius of only a few miles and is worthless. In the first there are no spark gaps and actions are enormously magnified by resonance. In both transmitter and receiver the currents are transformed and rendered more effective and suitable for the operation of any desired device. Properly constructed, my system is safe against static and other interference, and the amount of energy which may be transmitted is *billions of times greater* than with the Hertzian which has none of these virtues, has never been used successfully and of which no trace of which can be found at present.

A well advertised expert gave out a statement in 1899 that my apparatus did not work and that it would take 200 years before a message would be flashed across the Atlantic and he even accepted stolidly my congratulations on a supposed great feat. But subsequent examination of the records showed that my devices were

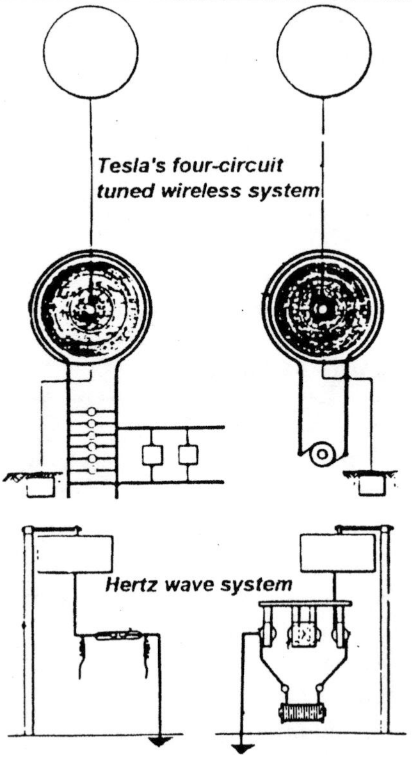

11. Tesla's 4-circuit system and Hertz's

secretly used all the time and ever since I learned of this I have treated these Borgia-Medici methods with the contempt in which they are held by all fair-minded men. The wholesale appropriation of my inventions was, however, not without a diverting side. As an example to the point I may mention my oscillation transformer operating with an air gap. This was in turn replaced by a carbon arc, quenched gap, an atmosphere of hydrogen, argon or helium, by a mechanical break with oppositely rotating members, a mercury interrupter or some kind of a vacuum bulb and by such *tours de force* as many new "systems" have been produced. I refer to this, of course, without the slightest ill-feeling, let us advance by all means. But I cannot help thinking how much better it would have been if the ingenious men, who have originated these "Systems," had invented something of their own instead of depending on me altogether.

Before 1900 two most valuable improvements were made. One of these was my individualized system with transmitters emitting a wave-complex and receivers comprising separate tuned elements cooperatively associated. The underlying principle can be explained in a few words. Suppose that there are *n* simple vibrations suitable for use in wireless transmission, the probability that any one will be struck by an extraneous disturbance is 1/n. There will then remain n - 1 vibrations and the chance that one of these will be excited is

$$\frac{1}{n - 1}$$

hence the probability that two tunes would be struck at the same time is

$$\frac{1}{n\,(n - 1)\,(n - 2)}$$

and so on. It will be readily seen that in this manner any desired degree of safety against statics or other kind of disturbance can be attained provided the receiving apparatus is so designed that its operation is possible through the joint action of all the tuned elements. This was a difficult problem which I have successfully solved so that now *any desired number of simultaneous messages is practicable in the transmission through earth as well as through artificial conductors.*

The other invention, of still greater importance, is a peculiar oscillator enabling the

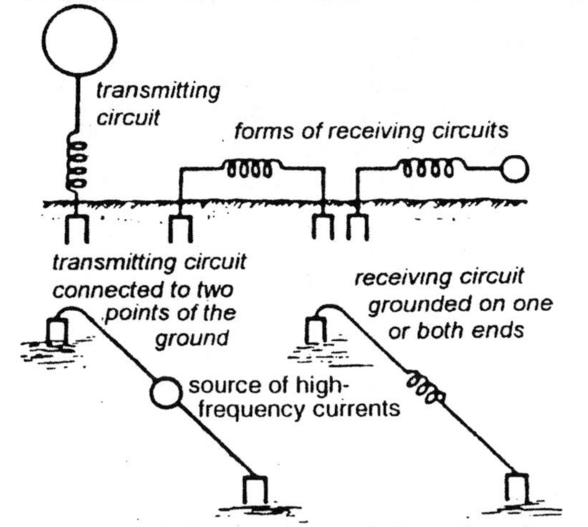

12. transmitting and receiving circuits

transmission of energy without wires in any quantity that may ever be required for industrial use to any distance and with very high economy. It was the outcome of years of systematic study and investigation, and wonders will be achieved by its means.

The prevailing misconception of the mechanism involved in the wireless transmission has been responsible for various unwarranted announcements which have misled the public and worked harm. By keeping steadily in mind that the transmission through the earth is in every respect identical to that through a straight wire, one will gain a clear understanding of the phenomena and will be able to judge correctly the merits of a new scheme. Without wishing to detract from the value of any plan that has been put forward I may say that they are devoid of novelty. So for instance in Fig. 12 arrangements of transmitting and receiving circuits are illustrated, which I have described in my U.S. Patent No. 613809 of November 8, 1898 on a Method of and Apparatus for Controlling Mechanism of Moving Vessels or Vehicles, and which have been recently dished up as original discoveries. In other patents and technical publications I have suggested conductors in the ground as one of the obvious modifications indicated in Fig. 5.

For the same reason the statics are still the bane of the wireless. There is about as much virtue in the remedies recently proposed as hair restorers. *A small and compact apparatus has been produced which does away entirely with this trouble*, at least in plants suitably remodeled.

Nothing is more important in the present phase of development of the wireless art than to dispose of the dominating erroneous ideas. With this object I shall advance a few arguments based on my own observations *which prove that Hertz waves have little to do with the results obtained even at small distances.*

In Fig. 13 a transmitter is shown radiating space waves of considerable frequency. It is generally believed that these waves pass along the earth's surface and thus affect the receivers. I can hardly think of anything more improbable than this "gliding wave" theory and the conception of the "guided wireless" which are contrary to all laws of action and reaction. Why should these disturbances cling to a conductor where they are counteracted by induced currents, when

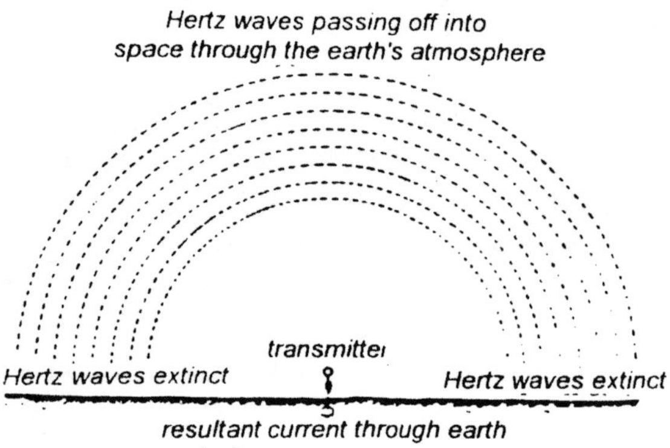

13. transmitter propagates in all directions

they can propagate in all other directions unimpeded?

The fact is that the radiations of the transmitter passing along the earth's surface are soon extinguished, the height of the inactive zone indicated in the diagram, being some function of the wave length, the bulk of the waves traversing freely the atmosphere. Terrestrial phenomena which I have noted conclusively show that there is no Heavyside layer, or, if it exists, it is of no effect. It would certainly be unfortunate if the human race were thus imprisoned and forever without power to reach out into the depths of space.

The actions at a distance cannot be proportionate to the height of the antenna and the current in same. I shall endeavor to make this clear by reference to the diagram in Fig. 14. The elevated terminal charged to a high potential induces an equal and opposite charge in the earth and there are Q lines giving an average current $=4Qn$ which circulates locally and is useless except that it adds to the momentum. A relatively small number of lines q, however, go off to a great distance and to these corresponds a mean current of $ie=4qn$, which is due to action at a distance. The total average current in an antenna is thus $Im=4qn$ and its intensity is no criterion for performance. The electrical efficiency of the antenna is

$$\frac{Q}{Q+q}$$

and this is often a very small fraction.

Dr. L.W. Austin and Mr. J.L. Hogan have made quantitative measurements which are valuable, but far from supporting the Hertz wave theory they are evidences in disproval of the same, as will be easily perceived by taking the above facts into consideration. Dr. Austin's researches are especially useful and instructive and I regret that I cannot agree with him on this subject. I do not think if his receiver was affected by Hertz waves he could ever establish such relations as he has found, but he would be likely to reach these results if the Hertz waves were in large part eliminated. At great distance the space waves and the current waves are of equal energy, the former being merely an accompanying manifestation of the latter in accordance with the fundamental teachings of Maxwell.

It occurs to me to ask the question: why have Hertz waves been reduced from the original frequencies to those I have advocated for my

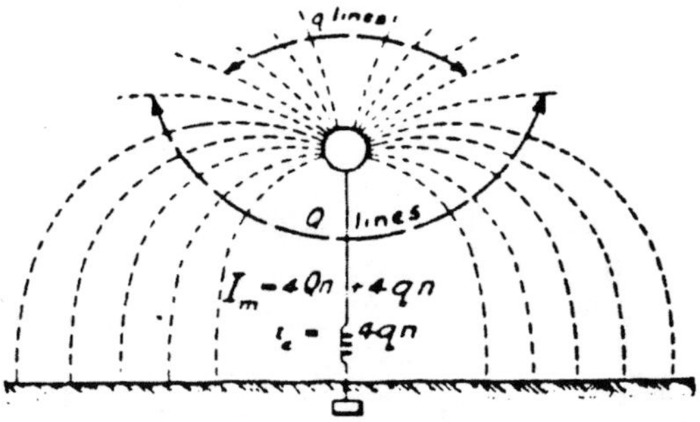

14 effective vs measured antenna current

system, when, in so doing, the activity of the transmitting apparatus has been reduced a billion fold? I can invite any expert to perform an experiment such as illustrated in Fig. 15, which shows the classical Hertz oscillator and my grounded transmitting circuit. It is a

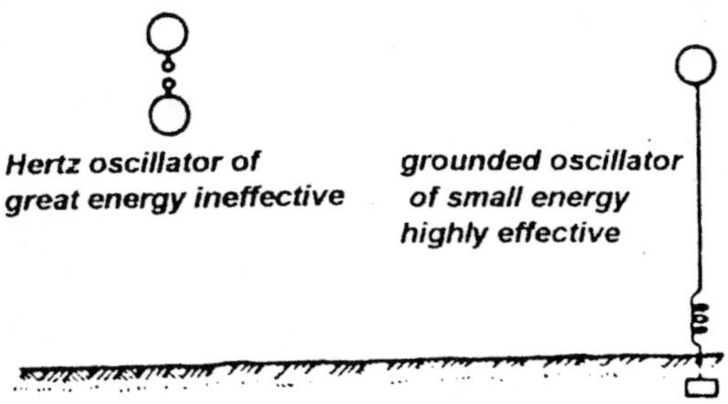

15. aerial vs. grounded radio

fact which I have demonstrated, that, although we may have in the Hertz oscillator an activity thousands of times greater, the effect on the receiver is not to be compared to that of the grounded circuit. This shows that *in the transmission from an airplane we are merely working through a condenser*, the capacity of which is a function of a logarithmic ratio between the length of the conductor and the distance from the ground. The receiver is affected in exactly the same manner as an ordinary transmitter, the only difference being that there is a certain modification of the action which can be predetermined from the electrical constants. It is not at all difficult to maintain communication between an airplane and a station on the ground, on the contrary, the feat is very easy.

To mention another experiment in support of my view, I may refer to Fig. 16 in which two grounded circuits are shown excited by oscillations of the Hertzian order. It will be found that the antennas can be put out of parallelism with

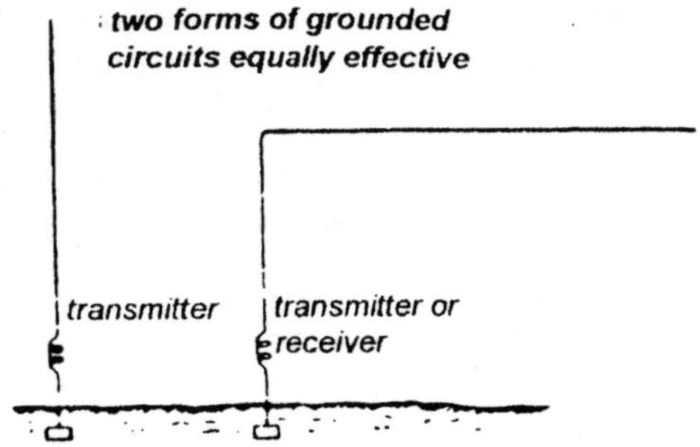

16. parallel antennas not necessary

out noticeable change in the action on the receiver, this proving that it is due to currents propagated through the ground and not to space waves.

Particularly significant are the results obtained

in cases illustrated in Figures 17 and 18. In the former an obstacle is shown in the path of the

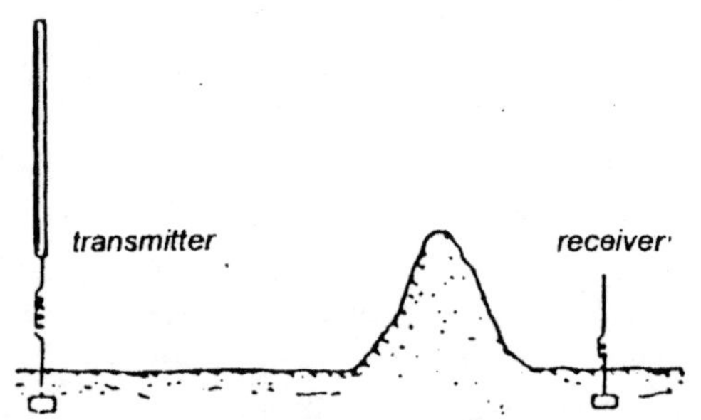

17. mountain between does not diminish

waves but unless the receiver is within the effective *electrostatic* influence of the mountain range, the signals are not appreciably weakened by the presence of the latter, because the currents pass under it and excite the circuit in the same way as if it were attached to an energized wire. If, as in Fig. 18, a second range happens to be beyond the receiver, it could only strengthen the Hertz wave effect by

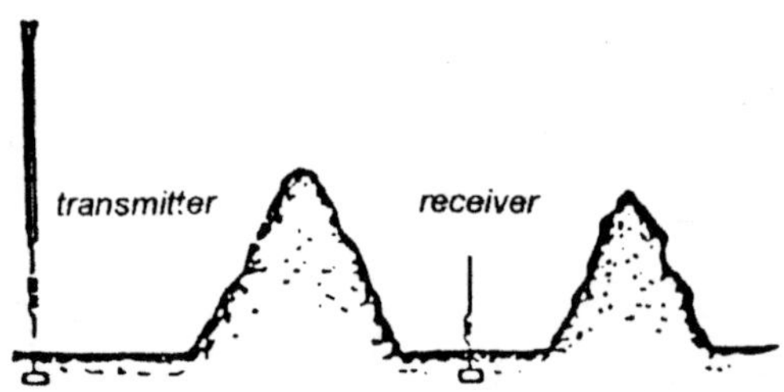

18. mountain behind does not reinforce

reflection, but as a matter of fact it detracts greatly from the intensity of the impulses Because the electric *niveau* between the mountains is raised, as I have explained in connection with my lightning protector in the Experimenter of February.

Again in Fig. 19 two transmitting circuits, one grounded directly and the other through an air

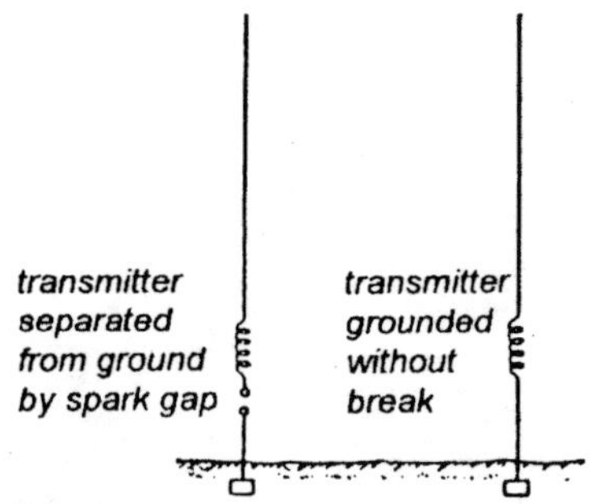

19. Hertz air gap not effective

gap, are shown. It is a common observation that the former is far more effective, which could not be the case in a transmission with Hertz radiations. In like manner if two grounded circuits are observed from day to day, the effect is found to increase with the dampness of the ground, and for the same reason also the transmission through sea water is more efficient.

An illuminating experiment is indicated in Fig.. 20 in which two grounded transmitters are shown, one with a large and the other with a

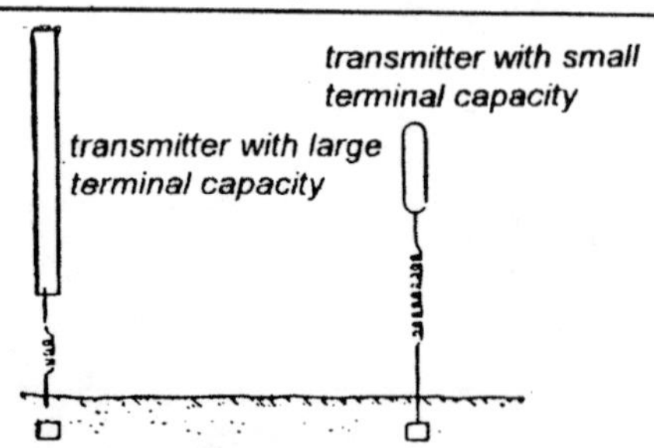

20. small aerial capacity not as effective

small terminal capacity. Suppose that the latter be one-tenth of the former but that it is charged to ten times the potential and let the frequency of the two circuits and therefore the currents in both antennas be exactly the same. The circuit with the smaller capacity will then have ten times the energy of the other, but the effects on the receiver will be in no wise proportionate.

The same conclusions will be reached by transmitting and receiving circuits with wires buried underground. In each case the actions carefully investigated will be found to be due to *earth currents*. Numerous other proofs might be cited which can easily be verified. So for example *oscillations of low frequency* are ever so much more effective in transmission, which is inconsistant with the prevailing idea. My observations in 1900 and the recent transmissions of signals to very great distances are another emphatic disproval.

The Hertz wave theory of wireless transmission may be kept up for a while, but I do not hesitate to say that in a short time it will be recognized as one of the most remarkable and inexplicable aberrations of the scientific mind which has ever been recorded in history.

Explanatory Notes

Maxwell's electromagnetic theory, pg. 1 James Clerk Maxwell (1831–79) developed the mathematics of electromagnetic phenomena and predicted that oscillations would produce wave propagations via an etheric medium, causing actions at a distance (i.e., radio). Maxwell identified the phenomenon with the physics of light, starting a tradition Tesla was to dispute.

Hertz's results, pg. 1 Heinrich Hertz (1857-94) conducted experiments with a spark-gap oscillator that appeared to confirm Maxwell's mathematical predictions of radio propagation. He, too, contended that radio behaved like light rays, having experimentally demonstrated reflection, etc. His ideas became fixtures in radio theory, and Hertz has been honored in recent years by renaming the unit of frequency after him (hz), although there is no law against using the old descriptive cycles per second (cps). Tesla contended that the waves Hertz described were not those that made radio work.

oscillation transformer, pg. 1 i.e., the Tesla coil, which became Tesla's potent transmitter.

Rhumkorf coil, pg. 1 a type of laboratory high-voltage induction coil that has a built-in vibrator enabling it to run on d.c.

1900 wireless transmitter, pg. 1 i.e., the magnifying transmitter Tesla developed during his Colorado Springs adventure of 1899-1900. Evolved from the Tesla coil and employing an extra third coil, this became Tesla's ultimate tool for radio and wireless power transmission.

heliographic, page 2 an old method of signaling that used a movable mirror to reflect beams of sunlight to a distance.

alternator, pgs. 2 and 3 an a.c. dynamo here cited by Tesla as his source of high-frequency oscillations when he lectured on the single-wire-without-return phenomenon. Tesla is on record as achieving 30 kc with such a device, which he also employed for experiments in high-frequency lighting. Tesla held many dynamo patents and was the consumate engineer in the field, but he abandoned the device because of its frequency limitations, employing instead his Tesla coil. The alternator did have a role in early low-frequency radio transmitting.

article in Century Magazine, pg 5, Tesla's article "On Increasing Human Energy" of June 1900, following his return from Colorado, was celebrated when it appeared, although, as an account of Tesla's Colorado work, it had been drastically edited.

rotating brush, pg. 6 a glowing discharge formed when high-voltage, high-frequency currents are applied to a terminal centered in a high-vacuum bulb. The discharge resolves into a ray so sensitive to etheric disturbances in its vicinity that it responds to a person approaching. Tesla expresses enthusiasm here and in other writings for its potential as a radio detector. Exactly how its sensitivity could be harnessed for this purpose is not clear, but in the same paragraph above he refers to a **high-speed photographic method,** and a year later in the *Electrical Review,* he refers to a similar device and to "a vacuum tube of high intensity which throws its light through a fine slot on a sensitive plate." In the same article he progresses to speculations on television, using that word, and says that this is a "subject to which I have devoted more than 25 years of close study." Imagine a Tesla TV in which a rotating-brush device both detects the signal and paints the picture in one simple, direct translation. Is this what he had in mind?

concatenated, tuned circuits, pg 6. Linked together, as in a chain. This basic invention of Tesla's was adopted by modern radio as indispensable, while his other basic precepts were ignored. Tesla invented more elaborate interlinkings, such as multiplexing.

farcical wireless suit pg. 7 probably Marconi Wireless Telegraph vs. U.S. Marconi patented Tesla's technology then challenged one of Tesla's fundamental patents (645,576, Fig. 11). Marconi had the wealth and clout to keep the farce going until 1943, when it was finally struck down by the U.S. Supreme court.

Borgia-Medici methods, pg. 8 ruthless and devious, as were some of the stratagems of these two ruling